A K.A.P.S. Guide to Preparing for Disasters

The College Edition

Copyright (c) 2022 by Joy Semien

All rights reserved. No part of this book may be reproduced, distributed, or transmitted in any form or by any means, including photocopying, recording, or other electronic or mechanical methods, without the prior written permission of the author, except in the case of brief quotations embodied in reviews and other specific non-commercial uses permitted by copyright law.

Publisher:
THGM Publishing
P.O. Box 562 Geismar, Louisiana 70734
https://www.thekapsdisasterhub.com/

Letter From the Author

Dear College Student,

Over the last several decades, there has been an increase in both natural and human-made hazards. These hazards have included hurricanes, flooding, oil spills, fires, chemical spills, and pandemics. Which has led to an increase in the loss of life and property.

For college students, the risk of being impacted by a hazardous event has increased. These events have led to displacement, disruption, academic challenges, and a host of mental health issues. There are few resources that specifically target the preparation of college students for hazardous events.

This book is designed to provide you as a college student with the tools you need to be prepared for a hazardous event. After reading and completing the activities in this book you should be well-equipped to respond to a hazardous event.

This book is divided into four sections: Knowledge, Attitudes, Preparedness, and Skills. The Knowledge and Attitude section highlights various types of disasters and how to properly respond in the event that a disaster affects you, as a college student. The Preparedness and Skills section provides templates and activities you can use to prepare yourself for a hazardous event.

Once you have completely worked through each section of this book I challenge you to continue reviewing the information with your roommates, friends, and colleagues on campus so that <u>you all are always prepared</u>. In the event that there is a disaster be sure to take this book with you as a guide to keep you equipped with the tools, you need to overcome the disaster.

Enjoy!
Joy Semien

About the Author

Joy Semien is an award-winning author, presenter, and research scientist. As a research scientist, she works to develop strategies to prepare communities for disasters.

Ms. Semien's interest in disaster research peaked at an early age living in a small rural community in Southeast, Louisiana. As a child, she witnessed chemical explosions and regularly breathed in chemicals that were released from neighboring chemical plants. She watched and listened to the stories of family members who had suffered from cancer and other health disparities as a direct result of chemical exposure.

This experience instilled in her the desire to want to further her education so she can make a difference in the world's most vulnerable communities. Joy holds a Bachelor of Science degree from Dillard University (2015) in Biology with a minor in Chemistry and a master's degree from Texas Southern University (2017) in Urban Planning and Environmental Policy.

After pursuing an advanced education she launched L.E.E.D. With Joy LLC in an effort to teach those working in communities how to Listen, Engage, Empower, and Drive Change (L.E.E.D.). Most recently Joy launched The K.A.P.S. Disaster hub as a subsidiary of L.E.E.D. With Joy L.L.C. Today Joy writes and develops training to prepare communities for disasters. This subsidiary seeks to encourage the growth in disaster Knowledge, Attitudes, Preparedness, and Skills (K.A.P.S.) across communities.

To learn more about Joy and L.E.E.D. With Joy, visit https://leedingwithjoy.com. To learn more about The K.A.P.S. Disaster Hub, visit the website www.thekapsdisasterhub.com, and follow @leedwithjoy & @thekapsdisasterhub on Instagram, Facebook, and YouTube.

Are You Wearing your Disaster K.A.P.S?

ARE YOU WEARING YOUR DISASTER K.A.P.S?

Disasters are often one of the most sudden and unexpected challenges anyone can go through, especially college students. However, If you are prepared the stress associated with the events can be somewhat alleviated.

In an effort to ensure that you are well-equipped for a hazard, I use the use acronym K.A.P.S. to represent four important terms you should know: (1) Knowledge, (2) Attitude, (3) Preparedness, and (4) Skills.

1. **Knowledge:** Understanding disasters and how to respond to each one
2. **Attitude:** The belief that the way we think or perceive things can influence our decisions
3. **Preparedness:** Taking steps to adequately prepare for a disaster
4. **Skill:** Practicing what you know

KNOWLEDGE

To be prepared for a disaster you must first understand which types of hazards can impact you as a college student. You can then take steps to effectively respond and potentially minimize your risk of injury and/or damage.

Important Definitions

Disaster: an incident that causes damage to human lives, infrastructure, the environment, and the economy as a result of a natural or manmade event. – National Red Cross and Red Crescent Federation

Hazard: A hazard is any source of potential damage, harm, or adverse health effects on something or someone. – OSHSA

Types of Hazards

Natural Event
- Flood
- Fire
- Tornado
- Windstorm
- Wildfires
- Droughts
- Earthquakes
- Hail

Manmade Event
- Industrial Explosion
- Chemical Leak
- Oil Spill
- Fire
- Blackout
- Pandemics
- Terrorist Attacks

KNOWLEDGE

The word <u>vulnerability</u> refers to any weakness, that you may possess that can increase your risk of injury or damage. The more vulnerable you are the more susceptible you are to a hazardous event.

What makes you vulnerable?

Where You live, work, go to school, and play can have a big impact on how vulnerable you are to hazards.

Whose Vulnerable?

Students
Elderly
Children
Caregivers
Illness/ Disability/ Immobility
Lack of Income/Savings
Lack of Disaster Training
Lack of Supplies
Lack of Friends/Families
Immigration Status
Minority Populations
Social Inequities
Proximity to the Coast
Proximity to Industrial Corridors

KNOWLEDGE

Elderly, children, students, medically ill, disabled, and/or low-income groups are most vulnerable to the impacts of hazards this is often because of things like physical immobility and the inability to access resources.

People who are low-income, lack education, and have yet to be trained in disaster preparedness can have a higher risk of vulnerability. This is because people with these characteristics may or may not have the knowledge or skills needed to respond to a hazard.

Minority groups and undocumented immigrants are also groups regularly impacted by hazards. As well as groups that live close to hazards (i.e. coastal areas & industrial corridors).

ATTITUDE

Preparing for a hazard is not just collecting supplies and making a plan. It is also about changing your attitude to ensure you as a college student is mentally equipped to handle a hazardous event. In many cases the way a person perceives a disaster will impact their stress levels, so learning how to **S.T.O.P.** the stress before it starts is important.

Addressing Misconceptions

The first step to developing a positive attitude toward a disaster is to break all misconceptions. You can do this by remembering to:

1. Never underestimate a hazardous event because it can quickly turn into a disaster.
2. Always obtain your facts from the National or Local Weather Stations.
3. Follow National/Local social media sites.
4. Approach every disaster well-prepared.
5. Always be ready to evacuate.

Changing Perceptions

Next, reevaluate how you perceive or think an event will occur. Remember no one event is the same. Just because you have experienced one type of event multiple times does not mean you will be okay during the next event. <u>Stay prepared and always be ready to evacuate!</u>

ATTITUDE

Managing Stress

Finally, learn how to control your stress level. If stress is not managed appropriately it can lead to panic and delayed responses. The more you know, the more prepared you are, and the stronger your social networks are the less stress you will endure if a hazardous event should occur.

When you are stressed remember to S.T.O.P.

Stop talking, panicking, and just breathe.
Think: about the facts, and assess the risks.
Organize: your facts and gather your belongings
Proceed: do what you believe is right.

Don't **React**, Just Breathe, & **Respond**

React: When we don't think about what to do before we make a decision this can cause us to panic.

Respond: Instead of acting without thinking take a second to gather yourself and your thoughts so you can decide how to proceed.

PREPAREDNESS

Preparing for a hazard is a "continuous cycle of planning, organizing, training, equipping, exercising, evaluating, and taking corrective action in an effort to ensure effective coordination during the incident response." - DHS, 2022.

Sheltering in Place

1. Pre-stock your dorm/apartment with water, can goods, toilet paper, etc.
2. Fill your car with extra gas.
3. Fill all bathtubs and other big containers with water.
4. Fasten down all loose material in and around the dorm/apartment.
5. Work with officials to board up windows and seal any holes to the outside of the dorm/apartment.

Evacuating

1. Fill all cars with extra gas.
2. Bring your completed disaster supply kit.
3. Before leaving turn off and unplug all appliances except the refrigerator and the freezer.
4. Clean out the refrigerator and the freezer before you leave.
5. Lift up all valuable items off of the floors.
6. Update roommates, friends, and family members on your evacuation plans.

HOW TO RESPOND TO A CHEMICAL EMERGENCY

A chemical emergency is the release of a hazardous chemical or substance. This can occur via air, water, or soil through leaks or explosions. The release has the potential to harm people's health.

How do you protect yourself and your roommates?

Ideally, You will Shelter in Place using the following steps:

- Go inside the closest building
- Close/ lock all of the doors and windows (use tape to seal windows and other holes)
- Place a towel under the windows and doors
- Turn off all fans, air conditioners, and heaters
- Turn on the Television/radio and tune into your local station to obtain updated alerts
- Be sure to subscribe to your on-campus and local emergency management list-serv and follow their social media accounts.
- Listen to a local Emergency Alert System (EAS) station for emergency instructions from the city, county, or state officials.
- Stay inside until you hear the "All Clear" siren or are otherwise notified via phone, radio, t.v. etc.

HOW TO RESPOND TO A CHEMICAL EMERGENCY

Evacuation Steps:

- Grab your disaster preparedness kit
- Grab your disaster plan
- Cover your nose and mouth with a towel or mask
- Get to the car as quickly as possible
- Ensure all windows are up, vents are closed, and the air conditioner is off!
- Drive to safety!

Need to Report a Chemical Emergency:
- 911
- National Response Center at 800-424-8802
- Poison Control Center (1-800-222-1222)

HOW TO RESPOND TO A HURRICANE

A Hurricane is a tropical cyclone, which forms over tropical or subtropical waters. Hurricane season is typically between June 1 and November 30.

Things to Know:
There are 5 categories of a hurricane:
- Category 5: Catastrophic Damage Wind: > 155MPH
- Category 4: Extreme Damage Wind 131-155 MPH
- Category 3: Extensive Damage Wind: 111-130 MPH
- Category 2: Moderate Damage Wind: 96-110 MPH
- Category 1: Minimal Damage Wind: 74 to 95 MPH

How do you protect yourself and your roommates?

Shelter in Place
- Pre-stock your dorm/apartment with water, can goods, toilet paper, etc.
- Fill your car with extra gas.
- Fill all bathtubs and other big containers with water.
- Fasten down all loose material in and around the dorm/apartment.
- Work with officials to board up windows and seal any holes to the outside of the dorm/apartment.

HOW TO RESPOND TO A HURRICANE

How do you protect yourself and your roommates?

Evacuate
- Fill all cars with extra gas.
- Bring your completed disaster supply kit.
- Before leaving turn off and unplug all appliances except the refrigerator and the freezer.
- Clean out the refrigerator and the freezer before you leave.
- Lift up all valuable items off of the floors.
- Update roommates, friends, and family members on your evacuation plans.

Important Reminders
- It is important to stay up to date with developing storms. Families can visit websites like www.nhc.noaa.gov, www.NOAA.gov, and adar.weather.gov to stay up to date.

HOW TO RESPOND TO A TORNADO

A tornado is "a narrow, violently rotating column of air that extends from the base of a thunderstorm to the ground." - FEMA, 2022. Tornado season is typically between March- July.

Things to Know:

Tornado Watch: Issued by the storm prediction center and represents a larger area. The notice is saying a tornado is possible.

Tornado Warning: Issued by the local weather center, it means a tornado has been spotted therefore take cover!

How do you protect yourself and your roommates?

Shelter in Place

- If in a car GET OUT and take shelter in a building or a low-lying area.
- If in a mobile home GET OUT! Take shelter in a building with a sturdy foundation.
- If in a building with cement foundation go into a room with no windows (hallway or closet) or the lowest area of the building.
- Place a mattress over you or another durable piece of furniture to resist debris.
- "Crouch down on your knees and protect your head with your arms."

HOW TO RESPOND TO FLOODING

Flooding is an overflow of water onto dry land. Usually the result of water filling creeks and/or river beds at a fast pace without warning. Flooding typically leads to flash floods which are extremely harmful, fast, and "unpredictable". - NOAA, 2022

Things to Know:

Flood Watch: Issued by the national weather service and represents a larger area. The notice is saying that flooding is possible. Expected to last 12 to 24 hours.

Flood Warning: Issued by the local weather center, it means flooding is occurring, therefore take shelter! Issued 24-60 hours before crest.

How do you protect yourself and your roommates?

- Always check the T.V. or Radio for flooding information
- Decide whether to Stay on campus or to Evacuate (If not mandatory)
- Make sure mobile or loose items are bolted to the ground & elevate all valuables
- Collect sandbags to place around homes
- Leave during daylight and check Evacuation Routes

HOW TO RESPOND TO WINTER WEATHER

Freezing indicates that ice up to 1/4 inch is expected in the area. The event may last a few hours or multiple days. – NWS, 2022

Be aware: Utilities and communication services are at risk of going out. Those at greatest risk are students, the elderly, children, sick individuals, and pets.

Things to Know:

Winter Storm Watch: Issued 12 to 48 hours before a winter storm. Possible blizzard, heavy snow, heavy freezing rain, or heavy sleet.

Winter Storm Warning: Issued 12 to 24 hours before a winter storm. Blizzard, heavy snow, heavy freezing rain, or heavy sleet is occurring.

Winter Weather Advisory: Issued to notify the public of unsafe weather conditions (i.e. snow, freezing rain, freezing drizzle, and sleet). Exercise extreme caution, otherwise hazard can lead to life-threatening conditions.

HOW TO RESPOND TO WINTER WEATHER

How do you protect yourself and your roommates?

Before the Storm

- Weatherize your dorm/apartment (insulate, caulk, add weather strips, and cover pipes
- Ensure there is a carbon monoxide detector installed
- Ensure there is a smoke detector installed
- Ensure there is a fire extinguisher available
- Check all heating devices for electrical shortages

During the Storm

Shelter in Place
- Avoid traveling
- Stay Indoors
- Drink warm Fluid
- Heat your dorm/apartment safely

Evacuation
- Dress warmly: cover all body parts fully to avoid frostbite
- Take your Disaster Kit & Plan

After the Storm

- Check news outlets for emergency information
- Check all pipes
- Check all heating equipment

HOW TO RESPOND TO SUMMER WEATHER

Extreme Heat indicates that the temperature will exceed 90 degrees and the humidity will become increasingly high for multiple days.

Be Aware: Utilities and communication services are at risk of going out. Those at greatest risk are the students, elderly, children, sick individuals, those overweight, and pets.

Things to Know:

Excessive Heat Watch: Issued by the local county. This indicates a heat index is possible.

Excessive Heat Warning: Issued by the local county, 1 to 2 days before a heat wave. Indicates a heat index of over 105 °F for at least two hours.

Excessive Heat Advisory: Issued by the local county, 1 to 2 days before a heat wave. Indicates a heat index of over 95°F for at least two hours.

HOW TO RESPOND TO SUMMER WEATHER

How do you protect yourself and your roommates?

Before the Heat Wave

- Weatherize your home (insulate, caulk, add weather strips, and install window reflectors)
- Locate your local cooling center
- Check all cooling devices for full operational capabilities and electrical shortages
- Install insulated window air conditioning Unit

During the Heat Wave

- Take cool showers or baths
- Stay inside cool doors
- Wear loose, lightweight, and light-colored clothing
- Cool your dorm safely
- Drink plenty of cool liquids
- Watch for heat-related illnesses
- Check on family/friends

HOW TO RESPOND TO A EARTHQUAKE

An Earthquake is when the ground suddenly begins to shake. This is typically caused by rocks under the earth's surface shifting. - USGS, 2022

The aftermath of an earthquake can lead to fires, tsunamis, landslides, and even avalanches.

How do you protect yourself and your roommates?

- Drop down to your hands and Knees
- Hold on to something sturdy
- Cover your neck and your head to prevent debris from falling on you
- If possible, get under a table or something sturdy
- If outside stay away from the building and brace yourself

SKILLS

Practice makes perfect! It is extremely important for you, as a college student to build a disaster preparedness kit and develop a plan! Once you have built the plan, then it is important to continuously practice the plan so you are prepared.

Steps to Build Your Skills

Design a plan

Build a disaster preparedness kit

Practice

Skills Section Instructions

The remainder of this book is designed to help you, as a college student build a disaster plan and a kit. As you build the plan don't feel intimidated, just breathe and take each section one step at a time.

Suggestions for Completion:

- <u>Host a preparedness party.</u> Invite your roommates, family, and colleagues over to work on the plan/kit together. Don't forget the music, food, and soft drinks.
- <u>Host a meditation gathering.</u> Create a safe space that allows everyone to have an open and honest conversation. Challenge attendees to share their thoughts and concerns.

Building

A Disaster Preparedness Kit

BUILDING A DISASTER PREPAREDNESS KIT

- [ ] Backpack
- [ ] Preparedness Plan
- [ ] Small Notebook
- [ ] Pens/Markers
- [ ] Emergency Contact Card
- [ ] Battery Packs/Batteries
- [ ] Chargers
- [ ] Candles/Matches
- [ ] Flashlight
- [ ] Earphones

THEKAPSDISASTERHUB.COM

BUILDING A DISASTER PREPAREDNESS KIT

- [ ] Non-perishable Food
- [ ] Disposable Utensils
- [ ] Disposable Bowls/Plates
- [ ] Manual can opener
- [ ] Water Purification Tablets
- [ ] 1 Gallon of Water - Per Person
- [ ] First Aid Kit (Band-Aids, Antibacterial ointments)
- [ ] Radio (Battery Operated)
- [ ] Tool kit (Pocket Knife, Screwdriver, Scissors etc.)
- [ ] Cleaning Supplies

THEKAPSDISASTERHUB.COM

BUILDING A DISASTER PREPAREDNESS KIT

- Extra Clothes & Shoes (3 days worth per person)
- Toiletries (Toothbrush, Soap, Deodorant)
- Hand sanitizer
- Sleeping bags, Blankets, & Pillows
- Medications
- Medical Equipment
- Towels
- Toilet paper / Paper Towels
- Ziplock Bags
- Garbage Bags

THEKAPSDISASTERHUB.COM

BUILDING A DISASTER PREPAREDNESS KIT

- [ ] Cell Phones
- [ ] Whistle
- [ ] Duct Tape
- [ ] Sunscreen/Bug Spray
- [ ] Umbrella/ Raincoat
- [ ] Particle Respirator/N95 Mask
- [ ] Disposable Camera
- [ ] Cash & Quarters
- [ ] Gloves
- [ ] Pet Supplies

THEKAPSDISASTERHUB.COM

BUILDING A DISASTER PREPAREDNESS KIT

- ☐ Homework/Class Assignments
- ☐ Laptops/Tablets
- ☐ _____
- ☐ _____
- ☐ _____
- ☐ _____
- ☐ _____
- ☐ _____
- ☐ _____
- ☐ _____

THEKAPSDISASTERHUB.COM

BUILDING A DISASTER PREPAREDNESS KIT

THEKAPSDISASTERHUB.COM

Notes

Let's Make a Plan!

Identification Cards

Using the space below tape/glue in pictures of your identification cards/ passport/green card/visa etc..

Emergency Contact Information

EMERGENCY HOTLINE

Mobile:_____

Telephone:_____

Email:_____

POISON CONTROL CENTER

Mobile:_____

Telephone:_____

Email:_____

HOSPITAL EMERGENCY

Mobile:_____

Telephone:_____

Email:_____

EMERGENCY CONTACT FAMILY

Mobile:_____

Telephone:_____

Email:_____

FIRE DEPARTMENT

Mobile:_____

Telephone:_____

Email:_____

POLICE DEPARTMENT

Mobile:_____

Telephone:_____

Email:_____

PHARMACY

Mobile:_____

Telephone:_____

Email:_____

EMERGENCY CONTACT FRIEND

Mobile:_____

Telephone:_____

Email:_____

Important Family Members

Using the space below record the contact information for all of your important family members who can serve as an emergency contact.

Name	Phone	Address	Relationship

Important Friends

Using the space below record the contact information for all of your important friends who can serve as an emergency contact.

Name	Phone	Address

On-Campus Resources

Using the space below record the on-campus resources that you will need during an emergency.

Name	Phone	Website	Resource Type

Off-Campus Resources

Using the space below record the off-campus resources that you will need during an emergency.

Name	Phone	Website	Resource Type

Notes

Be Prepared

Evacuation Information

EVACUATION CHECKLIST

- [ ] Emergency Contact Card
- [ ] Medications
- [ ] Social Security Card
- [ ] Passport/ID/Visa/Green card
- [ ] Copy of Birth Certificates
- [ ] Medical Information
- [ ] Pet Records
- [ ] Monthly Bills
- [ ] Bank Account Information
- [ ] Insurance Information (Health, Car, House)

EVACUATION CHECKLIST

☐ Fully Packed Disaster Preparedness Kit

☐ _____

☐ _____

☐ _____

☐ _____

☐ _____

☐ _____

☐ _____

☐ _____

☐ _____

THEKAPSDISASTERHUB.COM

THINGS TO REMEMBER

When Evacuating

1. Always check the T.V. or Radio for flooding information.
2. Decide whether to Shelter in Place or Evacuate (If not mandatory).
3. Make sure mobile or loose items are bolted to the ground.
4. Elevate all valuables.
5. Collect sandbags to place around dorms/apartments (if applicable).
6. Always check the T.V. or Radio for flooding information.
7. Leave during daylight and check evacuation routes.

When Returning Home

1. Wear protective clothing. ex. gloves, long sleeves, closed-toe shoes, and pants.
2. Check for damage to the dorm/apartment before entering.
3. Do not use an open flame for light.
4. Check food for spoilage.
5. Don't Drink the Tap Water.

Be Prepared

Medical Information

Doctor's Information

Using the space below record all relevant patient information.

Patient Name	Doctor's Name	Doctor's Phone #

Prescription Information

Using the space below record all relevant prescription information.

Patient Name	Medication Name	Doctor (Name & Phone #)	Pharmacy (Name & Phone #)

Medical Equipment

Using the space below record all relevant medical equipment information.

Patient Name	Equipment Name	Serial #	Company Contact (Name & Phone)

Medical Insurance Cards

Using the space below tape/glue in pictures of your insurance cards.

Perscription Cards

Using the space below tape/glue in pictures of your prescription cards.

Immunization Records

Using the space below tape/glue in pictures of your immunization records

Immunization Records

Using the space below tape/glue in pictures of your immunization records

Notes

Notes

Be Prepared

Insurance Information

Insurance Information

Using the space below record all relevant insurance information.

Insurance Type	Company Name	Policy #	Agent Contact Information

Insurance Cards

Using the space below tape/glue in pictures of your insurance cards.

Insurance Cards

Using the space below tape/glue in pictures of your insurance cards.

Notes

Notes

Be Prepared

Finance Information

Bank Information

Using the space below record all relevant bank information.

Bank Name	Phone # /Address	Account Type	Website	Username/ Password (Hint only)

Monthly Bills

Using the space below record all relevant monthly bills.

Bill	Company	Account #	Phone #	Due Date

Notes

Be Prepared

Vehicle Information

Vehicle Information

Using the space below record all relevant vehicle information.

Vehicle	VIN	Owner Name	Insurance Agency	Policy #

Vehicle Title

Using the space below tape/glue in pictures of your vehicle title.

Notes

Be Prepared

Pet Information

PET CHECKLIST

- Adoption Records and Registration for Each Pet
- Vaccination Records
- Medication Tracker
- Picture of You and Your Pet
- Food & Water
- Bowls
- Leashes
- Pet Carriers
- _____
- _____

PET CHECKLIST

THEKAPSDISASTERHUB.COM

Pets

Using the space below record all relevant pet information.

Pet Name	Vet Name	Vet Contact (Phone #)	Pet Shelter/ Boarding Facility

Pet Records

Using the space below tape/glue in pictures of your pet records.

Notes

Be Prepared

Education

High School Transcript

Using the space below tape/glue in pictures of your transcripts.

College Transcripts

Using the space below tape/glue in pictures of your transcripts.

Grad. School Transcripts

Using the space below tape/glue in pictures of your transcripts.

Current Class Information

Using the space below record all current course information and assignments you are responsible for completing if told to evacute.

Course	Teacher	Teacher E-mail	Assignment

Degree Plan

Using the space below tape/glue in a copy of your degree plan be sure to include course number, and descriptions.

Degree Plan

Using the space below tape/glue in a copy of your degree plan be sure to include course number, and descriptions.

Degree Plan

Using the space below tape/glue in a copy of your degree plan be sure to include course number, and descriptions.

Be Prepared

Dorm Inventory

Electronic Inventory

Using the space below make an inventory of all electronics in your home

Item	Serial Number	Manufacturer	Warranty Info.

Electronic Inventory

Using the space below make an inventory of all electronics in your home

Item	Serial Number	Manufacturer	Warranty Info.

Electronic Pictures

Using the space below tape/glue in pictures of your electronics.

Electronic Pictures

Using the space below tape/glue in pictures of your electronics.

Appliance Inventory

Using the space below make an inventory of all appliances' in your home.

Item	Serial Number	Manufacturer	Warranty Info.

Appliance Pictures

Using the space below tape/glue in pictures of your appliances.

Furniture Inventory

Using the space below make an inventory of all furniture in your home

Item	Serial Number	Manufacturer	Warranty Info.

Furniture Pictures

Using the space below tape/glue in pictures of your furniture.

What's in Your Fridge?

Using the space below make an inventory of the food in your fridge.

Item	Expiration Date

What's in Your Fridge?

Using the space below make an inventory of the food in your fridge.

Item	Expiration Date

Fridge Pictures

Using the space below tape/glue in pictures of the items you will be leaving in your fridge upon evacuation.

Helpful Hint:
Before leaving your house. Freeze a cup of water. Place a coin on top of the frozen water. When you return if the coin is below the level at which you left it you now know the power went out and you should dispose of all your food.

Fridge Pictures

Using the space below tape/glue in pictures of the items you will be leaving in your fridge upon evacuation.

What's in Your Freezer?

Using the space below make an inventory of the food in your freezer.

Item	Expiration Date

What's in Your Freezer?

Using the space below make an inventory of the food in your freezer.

Item	Expiration Date

Freezer Pictures

Using the space below tape/glue in pictures of the items you will be leaving in your freezer upon evacuation.

Helpful Hint:

Before leaving your house. Freeze a cup of water. Place a coin on top of the frozen water. When you return if the coin is below the level at which you left it you now know the power went out and you should dispose of all your food.

Freezer Pictures

Using the space below tape/glue in pictures of the items you will be leaving in your freezer upon evacuation.

Notes

Notes

Notes

Let's Practice

Scenario

Are you prepared?
Practice Scenario 1

It's 10:30 P.M. you and your roommates have been up all night studying for a chemistry midterm. All of a sudden you hear someone knocking on the door. Your roommate gets up to see who is knocking at your door. As you open the door your residential assistant begins to scream "The dorm is flooding! We have to evacuate now!" <u>What would you do next?</u>

1. What **type of disaster** is impacting you, as a college student?
   _____

2. Is this a **human-made** or **natural disaster**?
   _____

3. How would you **React** to this scenario? Why?
   _____
   _____

4. How would you **Respond** to this scenario? Why?
   _____
   _____

5. Do you **Evacuate** or do you **Shelter in Place**? Why?
   _____

6. What **items will you take with you**? Why?
   _____
   _____
   _____
   _____

7. Name ways to **stay calm** during this scenario?
   _____
   _____
   _____
   _____

Are you prepared?
Practice Scenario 2

It's 12:00 P.M. on a Friday you just left class. You and your classmates decide that you want to pick up some crawfish, snowballs, and go hang out by the lake. It's a beautiful day! After picking up the crawfish and the snowball you all turn the corner and realize the clouds are turning gray. So you decide to go back to the dorm. By the time you get to campus, the sky is so dark and the rain is so thick you can barely see. All of a sudden your cell phone begins to go off with weather alerts! The alert says "Take Cover Tornado Warning!" <u>What would you do next?</u>

1. What **type of disaster** is impacting you, as a college student?
   _____

2. Is this a **human-made** or **natural disaster**?
   _____

3. How would you **React** to this scenario? Why?
   _____
   _____

4. How would you **Respond** to this scenario? Why?
   _____
   _____

5. Do you **Evacuate** or do you **Shelter in Place**? Why?
   _____

6. What **items will you take with you**? Why?
   _____
   _____
   _____
   _____

7. Name ways to **stay calm** during this scenario?
   _____
   _____
   _____
   _____

Are you prepared?
Practice Scenario 3

It's 5:30 A.M. on a Wednesday – you were sleeping soundly. All of a sudden you smell smoke. You get up and look around but you don't see anything. You hop back in bed but the smell gets stronger so you nudge and yell to your roommate "wake up". In less than two minutes the whole dorm is covered in smoke. <u>What would you do next?</u>

1. What **type of disaster** is impacting you, as a college student?
   ----

2. Is this a **human-made** or **natural disaster**?
   ----

3. How would you **React** to this scenario? Why?

   ----

4. How would you **Respond** to this scenario? Why?

   ----

5. Do you **Evacuate** or do you **Shelter in Place**? Why?
   ----

6. What **items will you take with you**? Why?

   ----

7. Name ways to **stay calm** during this scenario?

Draw Your Evacuation Plan

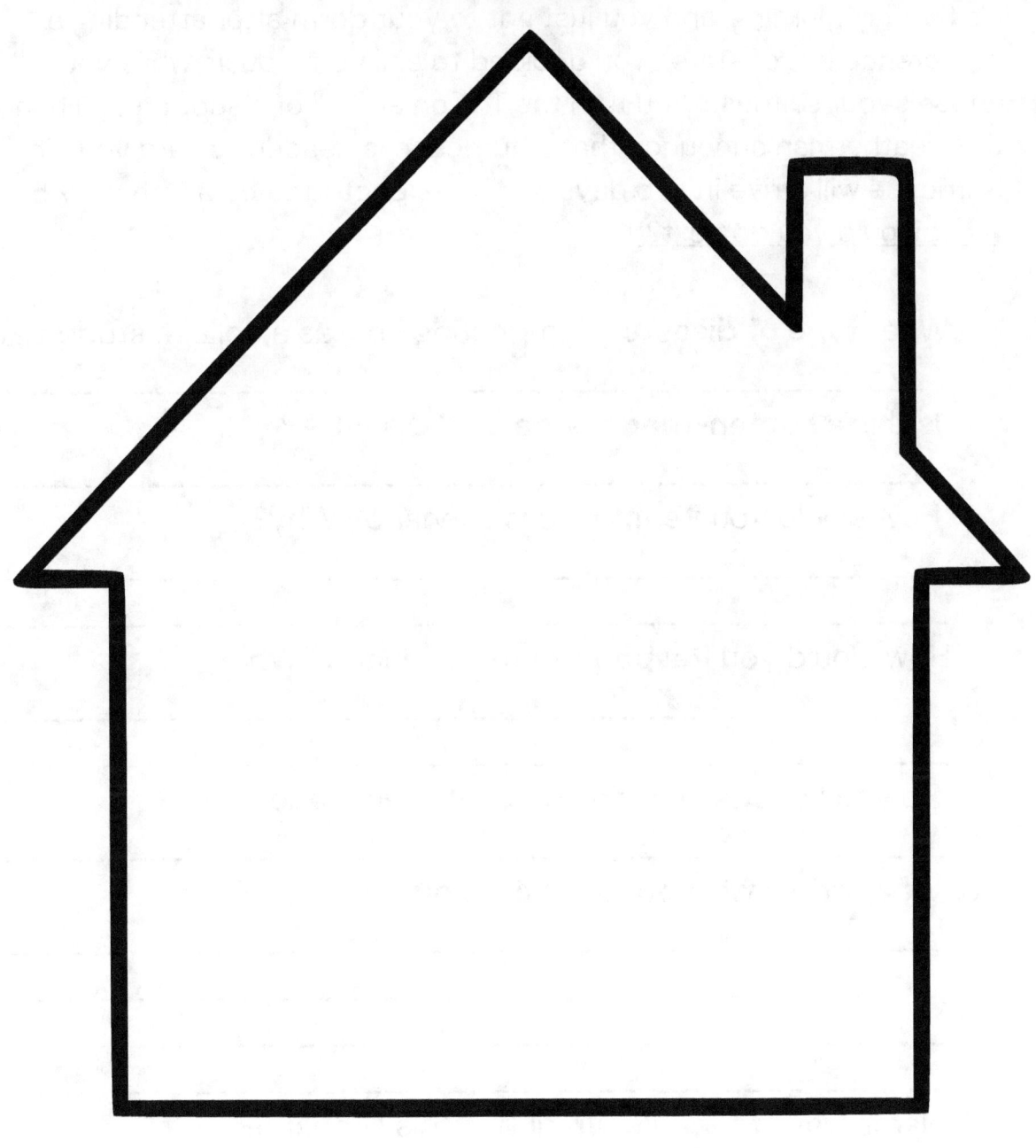

Are you prepared?
Practice Scenario 4

It's Sunday Morning, and you just got to your dorm after attending a conference out of state. You decided to call your cousin while you unpack your suitcase. You turn the T.V. on and all of a sudden you hear the weatherman announce that a hurricane is headed toward you. The Hurricane will arrive in two days and it's expected to be a Category 5. <u>What would you do next?</u>

1. What **type of disaster** is impacting you, as a college student?
   _____

2. Is this a **human-made** or **natural disaster**?
   _____

3. How would you **React** to this scenario? Why?
   _____
   _____

4. How would you **Respond** to this scenario? Why?
   _____
   _____

5. Do you **Evacuate** or do you **Shelter in Place**? Why?
   _____

6. What **items will you take with you**? Why?
   _____
   _____
   _____

7. Name ways to **stay calm** during this scenario?
   _____
   _____
   _____
   _____

Are you prepared?
Practice Scenario 5

It's Saturday morning, and you, are in the cafeteria sitting at the table eating Grits and Eggs for breakfast! You're enjoying your meal. All of a sudden you hear the chemical siren going off alerting you that there is a chemical emergency in the community. <u>What would you do next?</u>

1. What **type of disaster** is impacting you, as a college student?
   _____

2. Is this a **human-made** or **natural disaster**?
   _____

3. How would you **React** to this scenario? Why?
   _____
   _____

4. How would you **Respond** to this scenario? Why?
   _____
   _____

5. Do you **Evacuate** or do you **Shelter in Place**? Why?
   _____

6. What **items will you take with you**? Why?
   _____
   _____
   _____
   _____

7. Name ways to **stay calm** during this scenario?
   _____
   _____
   _____
   _____

Are you prepared?
Practice Scenario 6

It's Tuesday Afternoon, you are bored so you decided to watch the afternoon news. You hear the weatherman announce "Expect a winter freeze for the rest of the week. The roads will be covered in ice, snow will be falling, and there is a strong possibility that the power will be out." <u>What would you do next?</u>

1. What **type of disaster** is impacting you, as a college student?
   ----

2. Is this a **human-made** or **natural disaster**?
   ----

3. How would you **React** to this scenario? Why?

   ----

4. How would you **Respond** to this scenario? Why?

   ----

5. Do you **Evacuate** or do you **Shelter in Place**? Why?
   ----

6. What **items will you take with you**? Why?

   ----

7. Name ways to **stay calm** during this scenario?

Are you prepared?
Practice Scenario 7

It's Wednesday Morning, you are working out in the gym with the news blaring in the background. You hear the weatherman announce "Expect a major heat wave for your region over the next three days. There is a strong possibility of electrical power outages." <u>What would you do next?</u>

1. What **type of disaster** is impacting you, as a college student?
   _____

2. Is this a **human-made** or **natural disaster**?
   _____

3. How would you **React** to this scenario? Why?
   _____
   _____

4. How would you **Respond** to this scenario? Why?
   _____
   _____

5. Do you **Evacuate** or do you **Shelter in Place**? Why?
   _____

6. What **items will you take with you**? Why?
   _____
   _____
   _____
   _____

7. Name ways to **stay calm** during this scenario?
   _____
   _____
   _____
   _____

Are you prepared?
Practice Scenario 8

It's Sunday Morning, you are getting dressed for church. All of a sudden the dorm begins to shake you realize an earthquake is taking place. <u>What would you do next?</u>

1. What **type of disaster** is impacting you, as a college student?
   ----

2. Is this a **human-made** or **natural disaster**?
   ----

3. How would you **React** to this scenario? Why?

   ----

4. How would you **Respond** to this scenario? Why?

   ----

5. Do you **Evacuate** or do you **Shelter in Place**? Why?
   ----

6. What **items will you take with you**? Why?

   ----

7. Name ways to **stay calm** during this scenario?

Notes

Notes

Stay Prepared!

Remember:

Practice Makes Perfect! Keep this book updated and your family prepared!

References

Ascension Parish Emergency Preparedness Guide; www.ascension-caer.org

Bullard, R. D., Mohai, P., Saha, R., & Wright, B. (2008). Toxic wastes and race at twenty: why race still matters after all of these years. Envtl. L., 38, 371.

Bryce, Cyralene P. "Stress Management." *PAHO Library Cataloguing-in-Publication* (2001): 1-138. *PreventionWeb*. Pan American Health Organization. Web. 24 July 2016.

Cutter, Susan, Bryan J. Boruff, and W. Lynn Shirley. 2003. "Social Vulnerability to Environmental Hazards," Social Science Quarterly 84:2 (June 2003) pp. 242-261.

Capolla, Damon. Chapter 2: Preparedness. N.p.: FEMA: Emergency Management Institute, 4 Mar. 2005. DOC.

Chemical Emergencies." Emergency Preparedness and Response. Centers for Disease Control and Prevention, n.d. Web. 24 July 2016. <http://emergency.cdc.gov/chemical/overview.asp>.

Enayati, Amanda. "7 Ways to Manage Stress in a Disaster." CNN. Cable News Network, 31 Oct. 2012. Web. 23 July 2016. <http://www.cnn.com/2012/10/31/health/stress-disaster/>

FEMA. "Talking Points." America's PrepareAthon! Talking Points and Statistics(n.d.): n. pag. 5 Mar. 2014. Web. 23 July 2016.

Lee, Trymaine. "Cancer Alley: Big Industry Big Problem." MSNBC, n.d. Web. <http://www.msnbc.com/interactives/geography-of-poverty/se.html>.

"Natural Disasters." Ready.gov. Department of Homeland Security, n.d. Web. 05 July 2016.

NOAA. "Severe Weather 101." NOAA National Severe Storms Laboratory. The National Severe Storms Laboratory, n.d. Web. 23 July 2016.

"Office of Homeland Security and Emergency Preparedness." Ascension Parish. N.p., n.d. Web. 24 July 2016. <http://ascensionparish.net/>.

References

Paton, D., & Johnston, D. (2001). Disasters and communities: vulnerability, resilience and preparedness. Disaster Prevention and Management: An International Journal, 10(4), 270-277.

PTSD: National Center for PTSD." Traumatic Effects of Specific Types of Disasters -. USDVA, n.d. Web. 23 July 2016. <http://www.ptsd.va.gov/professional/trauma/disaster-terrorism/traumatic-effects-disasters.asp>.

"Preparing for a Disaster." International Federation of Red Cross and Red Crescent Societies. IFRC, n.d. Web. 24 July 2016. <http://www.ifrc.org/>.
"Preparedness Attitudes & Behaviors." Earth Institute. Columbia University. National Center for Disaster Preparedness, n.d. Web. 24 July 2016.
<http://ncdp.columbia.edu/research/preparedness-attitudes-behaviors/>.

"Preparedness Attitudes & Behaviors." Earth Institute. Columbia University. National Center for Disaster Preparedness, n.d. Web. 24 July 2016.
<http://ncdp.columbia.edu/research/preparedness-attitudes-behaviors/>.

Sutton, Jeannette, and Kathleen Tierney. (2006). *Disaster Preparedness: Concepts, Guidance, and Research*. Boulder: Natural Hazards Center Institute of Behavioral Science, Print.

Tornado Safety and How to Be Safe during a Tornado." Tornado Safety and How to Be Safe during a Tornado. Tornado Chaser, n.d. Web. 24 July 2016.
<http://www.tornadochaser.net/safety.html>.

Wingate, Martha S. et al. "Identifying and Protecting Vulnerable Populations in Public Health Emergencies: Addressing Gaps in Education and Training." *Public Health Reports* 122.3 (2007): 422–426. Print.

"Written Testimony of FEMA Administrator Craig Fugate for a House Committee on Transportation and Infrastructure, Subcommittee on Economic Development, Public Buildings and Emergency Management Hearing Titled "Blackout! Are We Prepared to Manage the Aftermath of a Cyber-Attack or Other Failure of the Electrical Grid?"" Homeland Security. N.p., 14 Apr. 2016. Web. 24 July 2016.
<https://www.dhs.gov/news/2016/04/14/written-testimony-fema-administrator-house-transportation-and-infrastructure>.

Other Resources

Check out these resources:
https://www.thekapsdisasterhub.com/
https://www.ready.gov/kit
https://www.ready.gov/publications
https://www.usgs.gov
https://hazards.fema.gov
https://www.nssl.noaa.gov/
https://www.redcross.org/
https://www.dhs.gov/prepare-my-family-disaster
https://youth.gov/
https://www.cdc.gov/cpr/readiness/resources.htm
https://echo.epa.gov/report-environmental-violations
https://988lifeline.org/

Footnotes

The materials in this book have been adapted from The Red Cross, The Federal Emergency of Management Agency, and Hazard Mitigation Training for Vulnerable Communities published by Routledge. This book should be used as a guide only. Be sure to connect with your local, regional, and national offices to obtain up-to-date resources when preparing for a disaster. All K.A.P.S. character images were created by Alexis Nunez while the the remainng images were downloaded from Canva stock images.

Other Books By Joy Semien

Community Development Books
Hazard Mitigation Training for Vulnerable Communities

Disaster Workbooks
A K.A.P.S. Guide to Preparing for Disasters: The Family Edition
A K.A.P.S. Guide to Preparing for Disasters: The Organization Edition
A K.A.P.S. Guide to Preparing for Disasters: Children's Edition
A K.A.P.S. Guide to Preparing for Disasters: College Edition
A K.A.P.S. Guide to Preparing for Disasters: College Edition Instructional Guide

www.ingramcontent.com/pod-product-compliance
Lightning Source LLC
Chambersburg PA
CBHW080501220526

45465CB00006B/2340